BAOFENG RADIO

Mastery Guide For Beginners

Logan Scott

TABLE OF CONTENTS

INTRODUCTION

INTRODUCTION

The Baofeng radio has emerged as a versatile and cost-effective solution the realm of emergency preparedness and survival communications. Baofeng, a Chinese manufacturer, has gained popularity for producing affordable and feature-rich handheld transceivers, making them a popular choice among preppers, outdoor enthusiasts, and emergency responders.

What is Baofeng Radio?

Baofeng radios, also known as Baofeng UV-5R radios, are compact, dual-band handheld transceivers that operate on VHF and UHF frequencies. These radios are designed to provide reliable communication over short to medium distances and come equipped with various features such as multiple frequency bands, programmable channels, and LCD displays. Baofeng radios are known for their affordability, making them accessible to a wide range of users.

Why Choose Baofeng for Survival Communications?

Affordability: One of the primary reasons for the widespread adoption of Baofeng radios is their affordability. In comparison to many other handheld transceivers on the market, Baofeng radios offer an attractive balance of features and cost, making them an excellent choice for budget-conscious individuals.

Versatility: Baofeng radios cover a broad spectrum of frequencies, enabling users to communicate on both VHF and UHF bands. This versatility is crucial in emergency situations where access to different frequency ranges might be necessary for effective communication.

Programmability: Baofeng radios are programmable, allowing users to customize channel settings, frequencies, and other parameters. This flexibility is advantageous for survival scenarios, as it enables users to adapt their radios to specific communication needs and changing conditions.

Compact and Portable: The compact and lightweight design of Baofeng radios makes them highly portable, facilitating easy carry during outdoor activities or emergency situations. Their handheld form factor enhances their suitability for on-the-go communication needs.

Long Battery Life: Baofeng radios typically feature long-lasting battery life, a crucial factor for survival situations where access to power sources may be limited. This ensures extended communication capabilities without the constant need for recharging.

Amateur Radio Community Support: Baofeng radios are popular among amateur radio operators, fostering a supportive community. This community-driven aspect provides users with valuable resources, such as programming guides, tips, and a wealth of shared knowledge.

In conclusion, Baofeng radios offer an affordable, versatile, and feature-rich solution for survival communications. Their compact design, programmability, and broad frequency coverage make them an attractive choice for individuals seeking reliable communication tools in various emergency scenarios.

Legal and Regulatory Considerations

When using Baofeng radios or any other radio communication equipment, it's crucial to be aware of legal and regulatory considerations to ensure compliance with relevant laws. Here are some key points to consider:

FCC Regulations:
In the United States, the Federal Communications Commission (FCC) regulates the use of radio frequencies. Baofeng radios must comply with FCC rules and regulations.

Baofeng radios are designed for Part 90 (Land Mobile) and Part 97 (Amateur Radio) services. Users should ensure they are operating within the appropriate frequency bands allocated for their specific use.

Amateur Radio Licensing:
If you are using Baofeng radios for amateur radio (ham radio) purposes, you must obtain an Amateur Radio License from the FCC. Different license classes allow access to various frequency bands and modes of communication.

Part 95 Certification:
Baofeng radios that operate on Family Radio Service (FRS) and General Mobile Radio Service (GMRS) frequencies should comply with Part 95 regulations. GMRS requires users to obtain a license from the FCC.

Use of Licensed Frequencies:
Baofeng radios should not be used on frequencies for which you do not have the proper license or authorization. Unauthorized use of specific frequencies can result in fines and penalties.

Power Limitations:
Baofeng radios, like other transceivers, are subject to power limitations set by regulatory authorities. Ensure that the power output of the radio complies with the maximum allowed limits for the specific frequency band.

Encryption Restrictions:
In many countries, the use of encryption on certain radio frequencies is restricted. Users should be aware of the limitations on encrypting communications and comply with applicable regulations.

Interference and Harmful Interference:
Users of Baofeng radios should avoid causing interference to other radio services. Harmful interference can result in regulatory action, fines, or loss of privileges.

Emergency Frequencies:

Certain frequencies are reserved for emergency and distress communications. Users should be aware of these frequencies and avoid transmitting on them unless in an emergency situation.

Local Regulations:

Radio frequency regulations may vary by country and locality. Users should familiarize themselves with local regulations and comply with all applicable laws.

Restricted Areas:

In some areas, the use of certain radio frequencies might be restricted or prohibited. Users should be aware of any local restrictions, especially in sensitive areas like airports or government facilities.

Chapter 1

Familiarizing Yourself With Baofeng Radios

Utilizing your Baofeng radio can be both exciting and empowering. This chapter is crafted to provide new users with a comprehensive introduction to unboxing, setting up, and getting acquainted with the fundamental features of their Baofeng radios. Whether you're a novice seeking guidance or a seasoned user brushing up on the basics, this step-by-step walkthrough ensures that you unlock the full potential of your new communication tool. Let's delve into the essential steps to make your initial experience with the Baofeng radio seamless and enjoyable.

Unboxing Your Baofeng Radio

Box Contents:
Carefully open the box and take note of its contents. Typically, you'll find the Baofeng radio unit, antenna, battery pack, belt clip, wrist strap, charger, earpiece, and user manual.

Inspecting the Radio:
Take a close look at the radio unit for any physical damage during shipping. Ensure all components are present and undamaged.

Getting Started with installation

Installing the Antenna:
Screw the included antenna onto the top of the radio. Ensure it is securely attached to optimize signal reception.

Installing the Battery Pack:
Locate the battery pack and slide it onto the back of the radio until you hear a click. Ensure the battery is securely attached.

Charging the Battery:

Connect the charger to the radio and a power source. Allow the battery to charge fully before use. Most Baofeng radios have a charging indicator that turns green when the battery is fully charged.

Powering On and Off

Turning On the Radio:

Locate the power/volume knob on the top of the radio. Turn it clockwise to switch on the device. You should hear a beep, and the LCD screen will illuminate.

Adjusting Volume:

Use the same knob to adjust the volume level to your preference. Turning it clockwise increases the volume, while counterclockwise decreases it.

Turning Off the Radio:

To power off the radio, press and hold the power/volume knob until you hear a beep, and the screen turns off.

Exploring the Display

LCD Screen:

Familiarize yourself with the LCD screen. It displays the current channel, frequency, battery level, and various icons indicating features like dual-watch or scanning.

Keypad:

Get to know the keypad, which allows you to input frequencies, navigate menus, and access various functions.

Basic Operation

Selecting Channels:

Baofeng radios often have multiple channels. Use the channel selector knob to scroll through available channels.

Adjusting Frequencies:

If you're using a specific frequency, use the keypad to input the desired frequency. Press the "MENU" button to confirm.

Transmitting and Receiving:

Press the "PTT" (Push-To-Talk) button on the side of the radio to transmit. Release it to listen. Always check your local regulations before transmitting.

Advanced Features

Scanning:

Experiment with scanning features to automatically search for active channels. Refer to the user manual for instructions on how to use scanning functions.

Dual Watch:

Explore the dual-watch feature, allowing you to monitor two channels simultaneously. This can be useful for monitoring emergency channels.

Programming Channels:

If you have specific frequencies to program, refer to the user manual or online resources for step-by-step programming instructions.

Basic Components and Buttons of Baofeng Radio

Baofeng radios are renowned for their versatility and functionality, but getting familiar with the basic components and buttons is essential for effective use. The following are the key elements of your Baofeng radio, empowering you to navigate its features confidently.

1. Antenna:

The antenna plays a crucial role in signal reception and transmission. Ensure it is securely attached to the top of your Baofeng radio. The antenna is usually removable for easy replacement or upgrades.

2. LCD Display:

The LCD screen provides vital information, such as the current channel, frequency, battery status, and icons indicating various functions. Understanding the display is essential for monitoring and adjusting settings.

3. Keypad:

Baofeng radios feature a keypad with numerical buttons, function keys, and a menu button. Familiarize yourself with the layout as you'll use it for inputting frequencies, accessing menu options, and navigating through different functions.

4. Power/Volume Knob:

Located on the top of the radio, this knob serves a dual purpose. Turning it clockwise powers on the radio and adjusts the volume. Press and hold it to turn the radio on or off.

5. PTT (Push-To-Talk) Button:

Positioned on the side of the radio, the PTT button is critical for transmitting your voice. Press and hold it while speaking, and release to listen. Ensure you are on the correct channel and frequency before transmitting.

6. Channel Selector Knob:

Use the channel selector knob to scroll through available channels. It's a fundamental component for quickly accessing different frequencies. Be aware of the channel you are on, especially in group communications.

7. Battery Pack:

The battery pack is specific to your Baofeng radio model. It is usually located on the back of the device and provides the power needed for operation. Ensure it is securely attached and fully charged for optimal performance.

8. Speaker and Microphone:

The speaker and microphone are located on the front of the radio. The microphone is often integrated into the radio's body. When using an external microphone or earpiece, make sure it is compatible with your specific Baofeng model.

9. Flashlight (Optional):

Some Baofeng models come with an integrated flashlight. If your radio has this feature, locate the button that activates the flashlight. It can be a useful tool in low-light conditions.

10. Accessory Ports:

Baofeng radios may have ports for additional accessories such as earpieces, microphones, or programming cables. Familiarize yourself with these ports for future expansions or modifications.

11. Belt Clip and Wrist Strap:

These accessories enhance the portability of your Baofeng radio. The belt clip allows you to attach the radio to your clothing or belt, while the wrist strap provides extra security during outdoor activities.

Refer to your user manual for model-specific details and explore advanced features as you grow more comfortable with your communication device. Whether you're a newcomer or an experienced user, mastering these elements ensures that you can leverage the full capabilities of your Baofeng radio in various situations.

Installing and Charging the Battery

Properly installing and charging the battery is a crucial initial step to ensure optimal performance from your Baofeng radio. Below is a step-by-step instructions to help you get started with confidence.

Installing the Battery:

Locate the Battery Pack:

Identify the battery pack, which is typically a rechargeable lithium-ion battery designed for your specific Baofeng radio model.

Align the Battery Contacts:
Examine the contacts on both the radio and the battery pack. Align the contacts on the battery pack with those on the radio, ensuring a snug fit.

Slide and Click:
Slide the battery pack onto the back of the radio, following the grooves or guides provided. Apply gentle pressure until you hear a distinct click, indicating that the battery is securely in place.

Check for Secure Attachment:
Confirm that the battery is securely attached to the radio. A properly installed battery ensures reliable power during operation.

Charging the Battery:

Locate the Charger:
Retrieve the charger that comes with your Baofeng radio. It usually consists of a cable and a charging base.

Connect the Charger:
Identify the charging port on your Baofeng radio. It is typically located on the side or bottom of the device. Connect the charger cable to the radio, ensuring a secure fit.

Connect to Power Source:
Plug the other end of the charger cable into a suitable power source. This can be a wall outlet or a USB power adapter. Ensure the power source is compatible with the specifications provided in your radio's manual.

Charging Indicator:
Many Baofeng radios have a charging indicator light on the charging base or the radio itself. The indicator may turn red or display another color to signify that the charging process has begun.

Monitor Charging Status:

Keep an eye on the charging indicator. Once the battery is fully charged, the indicator may turn green or display a different color. Refer to your user manual for specific charging times.

Disconnect and Test:

Once the battery is fully charged, disconnect the charger from both the radio and the power source. Turn on your Baofeng radio and verify that it powers up successfully.

Tips and Considerations:

First-Time Charging:

For new batteries, it's advisable to perform an initial full charge before using the radio extensively. Follow any specific recommendations provided in the user manual.

Avoid Overcharging:

While modern lithium-ion batteries have built-in protection, it's generally advisable to avoid overcharging. Disconnect the charger once the battery is fully charged.

Regular Maintenance:

Periodically check the condition of the battery and charging contacts. Clean them if necessary to ensure a reliable connection.

By following these steps, you've successfully installed and charged the battery for your Baofeng radio. A properly charged and securely attached battery ensures that your radio is ready for use in various situations. As you continue to explore the capabilities of your Baofeng radio, always refer to the user manual for model-specific guidelines and recommendations.

Attaching Antenna and Accessories

Congratulations on your Baofeng radio! To maximize its functionality, it's crucial to correctly attach the antenna and any additional accessories.

Attaching the Antenna:

Select the Correct Antenna:
Ensure you have the appropriate antenna for your Baofeng radio model. Different models may have specific antennas designed for optimal performance.

Identify the Antenna Connector:
Locate the antenna connector on the top of your Baofeng radio. It's usually a threaded connector, and the antenna should have a matching threaded base.

Screw On the Antenna:
Align the threads on the antenna base with those on the radio's antenna connector. Gently screw the antenna onto the connector in a clockwise direction. Be careful not to overtighten.

Hand-Tighten:
Hand-tighten the antenna until it is securely attached. Avoid using tools to tighten, as this can damage the threads.

Adjust the Angle (if applicable):
Some antennas may have an adjustable angle. Position the antenna for optimal reception based on your communication needs.
Attaching Accessories (e.g., Earpiece, Microphone):

Identify Accessory Ports:
Locate the accessory ports on the sides of your Baofeng radio. These ports are typically covered by protective flaps.

Select the Correct Accessories:
Choose the appropriate accessory for your needs, such as an earpiece, microphone, or programming cable. Ensure that the accessory is compatible with your specific Baofeng model.

Remove the Protective Flap:
Carefully open the protective flap covering the accessory port you intend to use. Some Baofeng radios may have separate ports for different accessories.

Insert the Accessory Connector:
Insert the accessory connector into the corresponding port. Align the pins on the connector with the receptacles in the port and gently push until it clicks into place.

Secure the Cable:
Secure the accessory cable along your clothing or gear to prevent it from getting tangled or caught during use.

Test the Accessories:
Turn on your Baofeng radio and test the accessories. Ensure that the earpiece provides clear audio, the microphone is transmitting correctly, or any other accessory is functioning as expected.

Additional Tips:

Proper Handling:
Handle the antenna and accessories with care to avoid any damage to the connectors or cables.

Regular Checks:
Periodically check the connections to ensure they remain secure. Loose connections can lead to reduced performance.

Accessory Compatibility:

Always use accessories that are designed for your specific Baofeng radio model. Check the user manual for compatibility information.

Chapter 2

Exploring Baofeng Radio Settings

In this chapter, we'll dive into the settings of your Baofeng radio. Whether you're new to these radios or looking to fine-tune your experience, we'll guide you through the various features and how to customize them. From programming channels to adjusting frequencies, get ready to discover the essentials for optimizing your Baofeng radio to meet your specific communication needs.

Baofeng radios offer a myriad of customizable features within their settings, catering to a diverse range of communication needs. Users can seamlessly program their preferred frequencies into designated channels, simplifying access to frequently used communication points.

Dual Watch and Dual Reception capabilities elevate situational awareness by enabling simultaneous monitoring of two channels or reception on two frequencies. The incorporation of VOX (Voice-Activated Transmission) facilitates hands-free communication, automatically transmitting when the user speaks.

CTCSS and DCS (Tone Squelch) technologies mitigate interference by allowing radios to respond solely to signals with the correct tone or code. Scan functions empower users to effortlessly identify active channels or specific programmed frequencies. The adjustable power output, ranging from low to high settings, provides flexibility for conserving battery life or extending communication range.

Additional features such as backlit displays, emergency alarm functions, and time-out timers further contribute to the versatility of Baofeng radios, making them adaptable tools for various communication scenarios. Users can tailor their radio experience through keypad locks, channel naming, voice prompts, and even PC programming options, enhancing overall usability and personalization.

Familiarity with these features ensures that Baofeng radios are not just communication devices but versatile tools that can be finely tuned to suit individual preferences and needs. Always consult the user manual for specific details pertaining to your Baofeng radio model.

Power On/Off and Volume Adjustment

Powering on and off, as well as adjusting the volume, are fundamental functions of any Baofeng radio. To turn on the radio, locate the power/volume knob on the top of the device and turn it clockwise. A distinctive beep and the illumination of the LCD screen indicate successful activation.

Conversely, to power off the radio, press and hold the same knob until you hear a beep and the screen turns off. The same knob serves a dual purpose for adjusting volume levels. Clockwise rotation increases the volume, while counterclockwise decreases it. Mastering these basic operations ensures the seamless activation, deactivation, and volume control of your Baofeng radio, laying the foundation for effective communication in various situations.

Programming Channels and Frequencies

Programming channels and frequencies on your Baofeng radio is a crucial skill for tailoring your communication experience. Begin by entering the menu system, typically accessible through a dedicated button.

Navigate to the channel or frequency programming option using the keypad. Input the desired frequency using the numeric keys, ensuring accuracy. To program a channel, assign a channel number to the frequency and save the configuration. Baofeng radios often allow users to customize channel names for easy identification.

Save your settings and exit the menu to activate the programmed channels. Regularly referring to the user manual for your specific Baofeng model ensures precision in programming, enabling you to swiftly and effectively access the frequencies essential for your communication needs.

Baofeng radios provide users with versatile options for channels and frequencies, allowing for effective communication across a range of scenarios. Understanding these elements is essential for maximizing the utility of your radio:

Frequency Bands:
Baofeng radios typically cover both VHF (Very High Frequency) and UHF (Ultra High Frequency) bands. This dual-band capability provides users with access to a wide spectrum of frequencies.

VHF and UHF Channels:
Within the VHF and UHF bands, Baofeng radios support various channels. Channels are predefined frequencies allocated for specific purposes, and users can program these channels for easy access.

Programmable Channels:
Baofeng radios allow users to program specific frequencies into channels for quick and convenient use. This feature is particularly valuable for organizing and swiftly switching between preferred frequencies.

Emergency Channels:
Some Baofeng radios come pre-programmed with emergency channels, often reserved for distress signals and emergency communications. Familiarize yourself with these channels for enhanced safety.

GMRS and FRS Channels:
If your Baofeng radio supports General Mobile Radio Service (GMRS) and Family Radio Service (FRS), it provides access to specific channels allocated for these services. GMRS may require a license for legal operation.

Weather Channels:
Baofeng radios often include NOAA weather channels, providing real-time weather updates and emergency alerts. These channels are crucial for staying informed about local weather conditions.

Frequency Bands for Radio Amateurs:
Baofeng radios are popular among amateur radio operators, and they cover a range of frequencies designated for amateur radio use. Users with an Amateur Radio License can access these bands for communication with other licensed operators.

Repeater Channels:
Repeaters extend the range of Baofeng radios by receiving and retransmitting signals. Users can program repeater frequencies and access dedicated repeater channels to enhance communication over longer distances.

Duplex and Simplex Operation:
Baofeng radios support both simplex (direct communication between radios on the same frequency) and duplex (communication through a repeater) operation. Understanding these modes is crucial for effective communication.

Scanning Frequencies:
Baofeng radios have scanning features that allow users to automatically search for active frequencies. This is beneficial for discovering new channels or monitoring different bands.

Frequency Steps:
Baofeng radios can typically adjust the frequency in predefined steps, such as 2.5 kHz or 5 kHz. Understanding and setting the appropriate frequency step is important for accurate tuning.

Privacy Tones (CTCSS/DCS):

Baofeng radios offer privacy tones, including Continuous Tone-Coded Squelch System (CTCSS) and Digital Coded Squelch (DCS), allowing users to filter out unwanted transmissions and reduce interference.

Setting Squelch and CTCSS/DCS Tones

Configuring squelch and CTCSS/DCS tones on your Baofeng radio is essential for optimizing communication clarity and minimizing interference. That is, it enhances communication quality by reducing noise and interference.

Setting Squelch:

Access the Menu:

Enter the menu system on your Baofeng radio using the designated button, usually labeled "MENU."

Locate Squelch Setting:

Navigate through the menu options to find the squelch setting. It may be labeled as "SQ" or "Squelch."

Adjust Squelch Level:

Once in the squelch menu, use the arrow keys or rotary knob to adjust the squelch level. Higher levels reduce background noise but may mute weak signals. Experiment to find the right balance for your environment.

Save Settings:

Save your squelch settings by following the instructions in the menu. This ensures your preferred squelch level is retained for future use.

Setting CTCSS/DCS Tones:

Access the Menu:

Enter the menu system on your Baofeng radio using the designated button.

Locate CTCSS/DCS Settings:
Find the menu option related to CTCSS (Continuous Tone-Coded Squelch System) and DCS (Digital Coded Squelch) tones. These settings are often labeled as "CTCSS" and "DCS."

Select the Desired Tone:
Choose the specific CTCSS or DCS tone you want to use. These tones help filter out unwanted signals and reduce interference on shared frequencies.

Enable/Disable:
Activate or deactivate CTCSS/DCS tones as needed. If communicating with others, ensure that everyone is using the same tones for successful reception.

Save Settings:
Save your CTCSS/DCS settings to ensure consistent communication. The process for saving settings may vary, so refer to your Baofeng radio's user manual for precise instructions.

Practical Tips:

Test in Different Environments:
Experiment with squelch and CTCSS/DCS settings in various environments to find the optimal configurations for your specific conditions.

Coordinate with Others:
If communicating with a group, coordinate and agree upon specific CTCSS/DCS tones to ensure seamless communication.

Refer to User Manual:
Always refer to your Baofeng radio's user manual for model-specific details and step-by-step instructions on adjusting squelch and CTCSS/DCS settings.

Memory Channels and Channel Scanning

Memory channels and channel scanning are indispensable features on Baofeng radios, offering users enhanced convenience and adaptability in their communication endeavors. Users can store frequently used frequencies into memory channels, allowing for quick and direct access without the need for manual input.

These memory channels can often be named for easy identification, and some Baofeng models enable users to program them efficiently via a computer. Memory channel scanning is a valuable function that permits users to scan through stored frequencies, providing a rapid overview of available options.

Additionally, Baofeng radios typically offer manual scanning options for real-time identification of active channels, programmed scans for focusing on specific areas of interest, and dual-watch scans for monitoring two channels simultaneously. The presence of priority scanning and time-based scanning options further enhances the scanning capabilities.

Users may also create customized scan lists, tailoring the scanning process to their specific needs. By organizing memory channels logically and regularly updating stored frequencies, users can ensure an efficient and tailored communication experience.

Mastering these functions not only simplifies the tuning process but also allows users to adapt their Baofeng radios seamlessly to different scenarios. As always, consulting the user manual specific to your Baofeng model provides valuable insights and instructions for optimal utilization.

Chapter 3

Making a Simple Call

Baofeng radios offer an accessible and reliable means of communication for a wide range of users, fostering connectivity in both recreational and professional settings. The simplicity of operation, affordability, and adaptability make Baofeng radios popular among diverse user groups.

Baofeng radios facilitate communication through a straightforward process that involves selecting frequencies, adjusting settings, and using the Push-To-Talk (PTT) button. Users can make calls using simplex or duplex modes, depending on the desired range and communication needs.

Key Steps in Making Baofeng Radio Calls:

Power On and Frequency Selection: Users power on the radio and choose the desired frequency manually or through preset channels.

Adjusting Settings: Power levels, CTCSS or DCS tones, and other settings are adjusted based on the communication requirements.

Antenna Attachment: Ensuring the antenna is securely attached optimizes signal transmission.

Listening Before Transmitting: Users listen to the channel to ensure it's clear before transmitting, preventing interference with ongoing conversations.

Push-To-Talk (PTT): The PTT button is pressed while speaking and released to listen, enabling a smooth exchange of messages.

Clear and Concise Communication: Effective communication involves clear and concise articulation of messages, allowing for efficient two-way exchanges.

Acknowledgment and Wait: After transmitting, users wait for acknowledgment from the receiving party before continuing the conversation.

Who Uses Baofeng Radios

Baofeng radios are versatile and cater to a diverse user base. Common users include:

Recreational Users: Enthusiasts engaging in outdoor activities like hiking, camping, or off-road adventures use Baofeng radios for group communication.

Emergency Preparedness: Baofeng radios serve as reliable communication tools during emergencies, allowing individuals to stay connected when other forms of communication may be compromised.

Volunteer Organizations: Various volunteer groups, such as community watch or event organizers, utilize Baofeng radios to coordinate activities efficiently.

Professionals: Baofeng radios are cost-effective options for professionals working in sectors like security, construction, or event management, where reliable communication is crucial.

Communication Dynamics:

Short-Range Communication: Baofeng radios excel in short to medium-range communication, making them ideal for group activities within a specific vicinity.

Direct and Immediate: The simplicity of Baofeng radios facilitates direct and immediate communication, promoting quick response times.

Simplex and Duplex Modes: Users can choose between simplex (direct communication) or duplex (using repeaters for extended range) based on their communication requirements.

In essence,

Understanding Call Signs and Etiquette

Call Sign:

A call sign is a unique identifier used in radio communication to distinguish individual users or groups. It serves as a personalized label that facilitates clear and efficient communication. Call signs are particularly important in scenarios where multiple users share the same frequency, ensuring each participant can be easily recognized. Call signs can be alphanumeric combinations or words and are crucial for establishing identity and orderliness on the airwaves. Examples of call signs are:

- **Individual Call Signs:** Alpha1, SierraEchoBravo, TangoLimaDelta, WhiskeyFoxTrot, BravoCharlie91, EchoRomeo123.
- **Group or Team Call Signs:** RedFoxTeam, MountainHikers, DeltaSquad, SearchRescue1, BravoEmergency, EchoResponse.

Etiquette in Radio Communication:

Radio etiquette refers to the set of practices and guidelines that individuals follow to ensure effective, respectful, and orderly communication over the airwaves. Adhering to radio etiquette is essential for maintaining clear channels, preventing interference, and fostering a positive communication environment. Key elements of radio etiquette include:

Listening Before Transmitting:

Before transmitting, individuals should listen to the channel to ensure it's clear. Avoid interrupting ongoing conversations.

Clear and Concise Messages:

Keep transmissions brief and to the point, conveying essential information without unnecessary chatter.

Identification:
Begin transmissions by identifying yourself or your group using your assigned call sign. This practice enhances clarity and accountability.

Pausing for Responses:
After transmitting, allow a brief pause to give others an opportunity to respond. Avoid consecutive transmissions without breaks.

Emergency Procedures:
In case of emergencies, follow established protocols, switch to designated emergency frequencies, and clearly communicate the nature of the emergency and your location.

Respecting Frequency Usage:
Respect specific frequencies for designated purposes (e.g., emergency channels) and comply with local regulations to avoid interference.

Repeater Courtesy:
When using repeaters, wait for a break in activity before transmitting. This ensures efficient use of the repeater system.

Consideration for Others:
Be mindful of other radio users, avoid monopolizing the channel, and consider the needs of other users sharing the frequency.

Avoiding Sensitive Information:
Refrain from transmitting sensitive or private information over open channels. Use secure communication methods when necessary.

Following Local Regulations:
Adhere to local radio regulations and licensing requirements to ensure legal and responsible use of radio frequencies.

Range and Limitations

The range and limitations of a Baofeng radio call, like any other radio communication, are influenced by various factors. Understanding these factors helps Baofeng radio users maximize their communication range while being aware of potential limitations. Regular testing in specific environments and conditions will provide a clearer understanding of the practical range for a given scenario.

1. Frequency Bands:
Baofeng radios commonly operate in VHF and UHF bands. VHF signals generally have better range in open spaces, while UHF signals penetrate obstacles better. The frequency band selected affects the effective range.

2. Power Output:
Baofeng radios often have adjustable power levels. Higher power settings extend the communication range, but they may consume more battery.

3. Antenna Quality and Gain:
The quality and type of antenna used play a crucial role. Upgrading to a higher-gain antenna can improve the radio's performance and extend the range.

4. Line of Sight:
Baofeng radios, like any other, rely on line-of-sight communication. Obstacles such as buildings, hills, or dense vegetation can limit the effective range by obstructing the signal path.

5. Repeater Usage:
Baofeng radios can use repeaters to extend their range. Utilizing repeater systems strategically placed in elevated locations enhances communication coverage, especially in challenging terrains.

6. Environmental Conditions:
Weather conditions, atmospheric disturbances, and geographical features can affect signal propagation. Rain, fog, or interference from other electronic devices may impact communication range.

7. Interference and Crowded Frequencies:
Baofeng radios may experience interference from other radio signals or electronic devices operating on the same frequencies. Using crowded frequencies can lead to reduced clarity and range.

8. Terrain and Urban Environments:
Baofeng radios may have limitations in urban environments with tall buildings and signal reflection. Signals may be absorbed or blocked, reducing effective range compared to open areas.

9. Radio Quality:
The performance of a Baofeng radio model may vary. Higher-quality models may have better sensitivity and signal processing capabilities, potentially improving range.

10. Regulatory Compliance: - Adhering to local radio frequency regulations is essential. Unauthorized use of frequencies or power levels may result in interference and legal consequences.

11. Battery Life: - Transmitting at higher power levels consumes more battery. Users should consider the balance between power levels and available battery life, especially in situations where extended communication is required.

12. User Experience: - The effective range of a Baofeng radio call is also influenced by the user's proficiency in optimizing settings, choosing appropriate frequencies, and employing good radio etiquette.

Chapter 4

Antenna and Signal Optimization

Antennae play a pivotal role in optimizing radio signal performance. A well-chosen antenna can enhance signal range and clarity. Consider factors like antenna type (whip, dipole), gain, and frequency compatibility. Optimal placement, such as raising the antenna and avoiding obstructions, further improves signal propagation. Regularly assess and maintain your antenna to ensure peak performance, contributing to effective communication in various scenarios.

Importance of Antenna Placement

The placement of the antenna on a Baofeng radio is of utmost importance for several reasons, influencing the overall performance of the radio communication. Efficient signal propagation relies on proper antenna placement, as it serves as a conduit for transmitting and receiving radio waves.

Signal Propagation:

Proper antenna placement ensures efficient signal propagation. The antenna acts as a conduit for transmitting and receiving radio waves, and its location can significantly impact the range and clarity of communication.

Line of Sight:

Radio signals generally operate in line-of-sight. Placing the antenna in a clear and elevated position helps minimize obstructions, maximizing the direct path between communicating radios and improving signal strength.

Avoiding Signal Blockage:

Placing the antenna away from obstructions such as buildings, trees, or other obstacles prevents signal blockage. Obstructions can attenuate or reflect radio waves, reducing the effective range of communication.

Reducing Interference:

A well-positioned antenna minimizes interference from nearby objects or electronic devices. Interference can distort signals and affect communication clarity, so optimal placement helps mitigate these issues.

Optimizing Radiation Pattern:

Antennas have a specific radiation pattern indicating the direction and strength of signal transmission. Proper placement aligns the radiation pattern with the intended communication direction, optimizing the effectiveness of the radio signal.

Maximizing Antenna Gain:

Antenna gain is a measure of its effectiveness in transmitting and receiving signals. Proper placement ensures that the antenna's gain is utilized to its fullest potential, enhancing the overall performance of the Baofeng radio.

Enhancing Range:

Effective antenna placement contributes to an extended communication range. Whether in urban or rural settings, placing the antenna for optimal performance is essential for reaching desired distances.

Improving Reception:

A well-placed antenna not only enhances transmission but also improves the radio's ability to receive signals. This is critical for receiving messages clearly, especially in environments with potential signal challenges.

Reducing Signal Loss:

Antenna placement minimizes signal loss during transmission. Placing the antenna in an optimal position helps ensure that the majority of the signal is transmitted and received, reducing the likelihood of communication errors.

In summary, antenna placement is a critical factor in optimizing the performance of a Baofeng radio. Whether for recreational use, emergency situations, or professional applications, proper antenna positioning enhances signal strength, range, and overall communication reliability. You should consider factors such as line of sight, obstruction avoidance, and elevation when determining the ideal placement for the antenna of your Baofeng radios.

External Antennas and Accessories

Baofeng radios are popular among amateur radio operators and users in various communication scenarios. Baofeng radios typically use handheld transceivers, and users may choose to enhance their performance with external antennas and accessories. Here are some common external antennas and accessories used with Baofeng radios:

External Antennas:

- **SMA-Female Antennas:** Baofeng radios often come with an SMA-Female antenna connector. External antennas compatible with this connector can be used to improve the radio's range and signal quality.
- **Whip Antennas:** These are flexible antennas suitable for handheld radios. They come in various lengths and may offer better performance than the stock antenna that comes with the Baofeng radio.
- **External Mobile Antennas:** For base station or mobile use, larger antennas with higher gain can be connected to Baofeng radios using appropriate adapters. These antennas are designed for mounting on a vehicle or a fixed location.

Antenna Adapters:

- **SMA-Male to BNC-Female Adapters:** Some users prefer using BNC connectors for external antennas. An adapter is needed to convert the SMA connector on the Baofeng radio to a BNC connector for compatibility with certain antennas.
- **SMA-Male to SO-239 Adapters:** For connecting Baofeng radios to external antennas with an SO-239 (UHF) connector, an adapter may be required.

Coaxial Cables:

- **Extension Cables:** Users may need extension cables to connect Baofeng radios to external antennas placed at a distance. Ensure the cable has the appropriate connectors on both ends.

Antenna Mounts:

- **Magnetic Mounts:** For mobile applications, a magnetic mount with an attached antenna can be placed on the roof of a vehicle to improve signal reception.

Speaker Microphones (Speaker Mics):

- **External Microphones:** These accessories provide a hands-free operation option and are useful when the radio is attached to a belt or carried in a pocket.

Always ensure that the external antennas and accessories you choose are compatible with the specific model of Baofeng radio you own, and consider factors like frequency compatibility, connector types, and the intended use case to optimize performance. Additionally, be aware of any legal and regulatory considerations related to antenna modifications and usage.

Signal Boosting Techniques

Improving the signal performance of a Baofeng radio involves various techniques and accessories. Few of the techniques are?

Use an External Antenna:
Replace the stock antenna with a higher-gain external antenna. Omni-directional or directional antennas can significantly improve the range and signal quality.

Choose the Right Frequency and Band:
Ensure that you are operating on the appropriate frequency and band for your specific communication needs. Baofeng radios cover multiple frequency bands, so select the one that suits your application.

Elevate the Antenna:
Positioning the antenna higher, such as on a rooftop or using a mast, can improve line-of-sight and signal propagation. This is especially useful in outdoor or open areas.

Utilize Repeaters:
Baofeng radios can benefit from using repeaters. Repeaters receive a weak signal and retransmit it at a higher power, extending the communication range. Make sure you have the necessary access permissions to use local repeaters.

External Antenna Cable:
Use low-loss coaxial cables for connecting external antennas. High-quality cables minimize signal loss over longer distances, ensuring that more of the transmitted power reaches the antenna.

Antenna Tuning:
Adjust the length of a wire antenna or a dipole antenna to the appropriate frequency. This tuning process can optimize the antenna's performance for the desired frequency.

Antenna Grounding:
Properly ground the external antenna to reduce interference and improve overall performance. Grounding can help protect against static discharges and improve signal quality.

Signal Boosters and Amplifiers:
Consider using signal boosters or amplifiers designed for handheld radios. These devices amplify the signal output, providing a stronger signal to the antenna.

Radio Placement:
Experiment with the placement of the radio. Avoid obstacles, such as buildings or large metal structures, that may obstruct the signal. Elevating the radio can also improve signal reception.

Use High-Capacity Batteries:
Ensure your radio has a fully charged, high-capacity battery. A weak battery can affect the overall performance of the radio, including its signal transmission capabilities.

Minimize Interference:
Identify and minimize sources of interference, such as electronic devices or other radios operating on the same frequency. Interference can degrade signal quality.

Always be mindful of legal and regulatory requirements when modifying or enhancing radio equipment. Additionally, be considerate of the frequencies you are using to avoid interference with other users.

Chapter 5

Troubleshooting and Maintenance

When it comes to ensuring the reliable performance of your Baofeng radio, understanding basic troubleshooting and maintenance practices is crucial. If you encounter issues like the radio not powering on, poor audio quality, or interference, start by checking the battery, cleaning contacts, and adjusting the antenna.

Ensure proper programming and be aware of your operating environment to avoid potential interference. Regularly inspect the antenna for damage and keep battery contacts clean. Additionally, maintain your batteries by charging and discharging them regularly. Protect your radio from water exposure, and consider firmware updates if applicable.

Periodic testing and familiarizing yourself with these troubleshooting and maintenance practices will go a long way in keeping your Baofeng radio operating at its best, ensuring effective and reliable communication when you need it most. Refer to the user manual for specific guidance tailored to your radio model.

Common Issues and Solutions

Baofeng radios are popular handheld transceivers known for their affordability and versatility. However, like any electronic device, they may encounter common issues. Here are some common problems associated with Baofeng radios and possible solutions:

Poor Battery Life:

Solution: Ensure you are using a fully charged and compatible battery. If the battery is old or damaged, consider replacing it. Avoid leaving the radio on when not in use, as this can quickly drain the battery.

Programming Difficulties:

Solution: Programming Baofeng radios manually can be challenging. Consider using programming software and a programming cable to simplify the process. Ensure you are entering frequencies, tones, and other settings correctly.

Noisy Transmission or Reception:

Solution: Check for obstructions or interference in your environment. Ensure that the antenna is properly connected and not damaged. Try changing your location or adjusting the squelch setting to filter out background noise.

Interference from Other Radios:

Solution: Change your operating frequency or use a different channel to avoid interference. If you are operating on a shared frequency, consider coordinating with other users to minimize conflicts.

Display Issues:

Solution: If the display is not working or showing garbled information, check the battery connection and try a different battery. If the issue persists, there may be a problem with the radio's internal components, and professional repair may be needed.

Low Transmit Power:

Solution: Ensure that the radio is set to the highest power setting. If the issue persists, the battery may be low or faulty. Check the antenna for damage, as a damaged antenna can reduce transmission efficiency.

Unintended Keypad Presses:

Solution: Be mindful of how you handle the radio to avoid inadvertently pressing buttons. Consider using a keypad lock function if available on your model.

Radio Won't Turn On:

Solution: Check the battery to ensure it is properly charged and connected. If the battery is good, inspect the power button for any physical damage. If the issue persists, there may be a problem with the internal circuitry, and professional repair may be necessary.

Inability to Access Certain Frequencies:

Solution: Check the frequency range limitations of your specific Baofeng model. Some frequencies may require special programming or may be restricted by local regulations.

Strange Noises During Transmission:

Solution: Ensure that the microphone is securely connected. If using an external microphone, check its condition. Strange noises may also be caused by interference or a weak signal. Adjust your location or try a different frequency.

If you encounter persistent issues with your Baofeng radio, it's advisable to consult the user manual or contact the manufacturer's customer support for further assistance.

Performing Regular Maintenance

Performing regular maintenance on your Baofeng radio can help ensure its optimal performance and longevity.

- **Inspect the Exterior:**

Regularly check the external components, such as the antenna, battery, and connectors, for any signs of damage or wear. Replace any damaged parts promptly.

- **Clean the Radio:**

Use a soft, lint-free cloth to clean the exterior of the radio. Avoid using abrasive materials or harsh chemicals, as they may damage the finish.

- **Check the Battery Contacts:**

Inspect the battery contacts for any corrosion or debris. Clean the contacts with a pencil eraser or a small brush to ensure a good connection between the battery and the radio.

- **Verify Antenna Connection:**

Ensure that the antenna is securely connected to the radio. Check for any loose connections or damage to the antenna. A damaged or improperly connected antenna can impact signal transmission.

- **Battery Maintenance:**

Charge rechargeable batteries before they are fully depleted to maximize their lifespan. If using rechargeable batteries, follow the manufacturer's recommendations for charging and storage.

- **Update Firmware:**

Check for firmware updates on the Baofeng website or through official channels. Updating the firmware can improve performance and address any known issues.

- **Perform Radio Checks:**

Regularly test the radio's functionality by performing radio checks with other users. Ensure that both transmission and reception are clear and that all features are working as expected.

- **Store Properly:**

When not in use, store your Baofeng radio in a cool, dry place. Avoid exposing it to extreme temperatures, humidity, or direct sunlight. Consider using a protective case to prevent physical damage during storage or transportation.

- **Avoid Overcharging:**

If using rechargeable batteries, avoid overcharging, as it can negatively impact battery life. Follow the recommended charging times provided by the manufacturer.

- **Inspect Accessories:**

Check any accessories such as microphones, earpieces, or programming cables for damage. Replace any worn-out or damaged accessories to maintain reliable communication.

- **Perform a Function Test:**

Regularly perform a function test to ensure all buttons, knobs, and features are working correctly. Address any issues promptly to prevent further damage.

- **Follow User Manual Guidelines:**

Refer to the user manual for your specific Baofeng model and follow any recommended maintenance procedures outlined by the manufacturer.

Chapter 6

Integrating Baofeng Radios with Other Survival Gear

In the face of unexpected emergencies or challenging situations, reliable communication can be a lifesaver. Baofeng radios, known for their affordability, versatility, and durability, have become a popular choice for preppers, survivalists, and outdoor enthusiasts. But their true potential shines when seamlessly integrated with other essential survival gear.

1. Power Up:

- Backup Batteries: Standard Baofeng batteries may not last long in demanding situations. Invest in extended-life batteries and keep them charged. Consider solar chargers or hand cranks for extended power needs.

- Portable Power Banks: A power bank can be a lifesaver for recharging your radio and other devices on the go. Choose one with sufficient capacity and compatible charging ports.

2. Enhance Communication:

- Headsets and Earpieces: Free your hands and improve audio clarity with a comfortable headset or earpiece. They're especially useful in noisy environments or when multitasking.

- External Antennas: Upgrade your signal range and reception with a detachable antenna. Choose one based on your desired frequency bands and operating environment.

3. Stay Informed and Entertained:
- **NOAA Weather Radio:** Most Baofeng models can receive NOAA weather broadcasts, providing crucial updates on storms, floods, and other hazards.

- **FM Radio:** Tune into local radio stations for news, information, and even entertainment during downtime.

4. Signaling and Navigation:

- GPS Devices: Pair your Baofeng with a GPS device for enhanced navigation and emergency beacon capabilities. Some models even offer direct communication with search and rescue teams.

- Signal Mirrors: Use a signal mirror to reflect sunlight and attract attention over long distances, complementing your radio communication efforts.

5. Additional Gear:

- Carrying Cases: Protect your Baofeng radio from the elements and accidental damage with a sturdy carrying case. Choose one with compartments for accessories and additional gear.

- Molle Pouches: Mount your radio and accessories onto your backpack or belt with Molle pouches for easy access and hands-free operation.

ALSO NOTE

Licensing: Depending on your location and intended use, you may need a license to operate a Baofeng radio. Familiarize yourself with local regulations before using your device.

Practice and Training: Learn how to use your Baofeng radio effectively before relying on it in critical situations. Practice basic communication protocols and emergency procedures.

Maintenance: Keep your Baofeng radio clean, dry, and well-maintained. Regularly inspect antennas, connections, and batteries to ensure optimal performance.

By integrating your Baofeng radio with other essential survival gear, you create a robust and versatile communication system that can be invaluable in unexpected situations. **Remember, knowledge is power, so invest time in learning and practicing to maximize the potential of your radio and stay connected when it matters most.**

Connecting Radios to GPS Devices

The process of connecting your Baofeng radio to a GPS device depends on the specific equipment you have and the desired functionality. Here's a breakdown of the two main approaches:

1. Connecting for APRS (Automatic Packet Reporting System):
This method allows you to broadcast your location and receive location data from other users on the same frequency.

Step 1: Gather your equipment:
- Baofeng radio with APRS capability (e.g., UV-5R, DMR-6X2)
- GPS device with NMEA 0183 output (e.g., Garmin GPSMAP 66i, Byonics TinyTrak4)
- TNC (Terminal Node Controller) or Bluetooth APRS app (e.g., APRSdroid, PocketPacket)
- Audio cable (TRS to stereo TRS or stereo TRS to stereo TRS)

Step 2: Connect the devices:
- Connect the GPS device's NMEA 0183 output to the TNC or Bluetooth adapter's input using the appropriate cable.
- Connect the TNC or Bluetooth adapter's audio output to the Baofeng radio's microphone input using the appropriate cable.

Step 3: Configure your Baofeng radio:
- Enable APRS functionality in your radio's settings.
- Set the correct APRS parameters like callsign, beacon interval, and channel.
- Refer to your Baofeng radio's manual for specific instructions.

Step 4: Configure your TNC or Bluetooth app:
- Set the correct baud rate and data format to match your GPS device's NMEA output.
- Configure the callsign, beacon interval, and channel to match your Baofeng radio settings.
- Refer to your TNC or app's manual for specific instructions.

Step 5: Test and verify:
- Turn on all devices and wait for the GPS signal to lock.
- Check if your Baofeng radio transmits your location data via APRS.
- If not, troubleshoot the connection and settings based on your equipment's manuals.

2. Connecting for Navigation and Emergency Beacons:
Some GPS devices offer direct communication with your Baofeng radio without needing a TNC. This allows you to send location data, trigger emergency beacons, and receive text messages.

Step 1: Check compatibility:
- Ensure your Baofeng radio and GPS device support direct communication features like Bluetooth or GMRS.
- Refer to your devices' manuals for compatible models and features.

Step 2: Pair the devices:
- Activate Bluetooth or GMRS pairing mode on both your Baofeng radio and GPS device.
- Follow the on-screen instructions for each device to establish the connection.

Step 3: Configure communication settings:
- Set up the communication channels and data formats on both devices.
- Refer to your devices' manuals for specific instructions.

Step 4: Test and verify:
- Send a test message or location update from your GPS device to your Baofeng radio.
- Ensure the communication works as expected.

Additional Tips:
- Choose a reliable TNC or Bluetooth app with good reviews and user support.
- Practice using your APRS setup and communication features before relying on them in critical situations.
- Always follow local regulations and licensing requirements for radio operation.

Utilizing Radios in a Bug-Out Scenario

"Bug out" is a term that originated from military slang and is commonly used in emergency preparedness and survivalist communities. It refers to the act of quickly leaving or evacuating a location due to an impending threat, disaster, or emergency situation. The goal of "bugging out" is to move to a safer location, often a pre-planned and prepared shelter or destination. The term is often associated with scenarios such as:

- **Natural Disasters:** Evacuating an area threatened by hurricanes, floods, wildfires, earthquakes, or other natural disasters.

- **Civil Unrest:** Leaving an area with the potential for civil unrest, social upheaval, or public safety concerns.

- **Pandemics:** Evacuating an area during the spread of contagious diseases or pandemics.

- **Terrorist Threats:** Moving away from an area facing immediate or potential terrorist threats.

- **Personal Safety Concerns:** Leaving a location due to personal safety concerns, such as a home invasion or imminent danger.

The concept of "bugging out" is closely tied to the idea of emergency preparedness and having a well-thought-out plan for responding to various threats. This plan typically includes considerations for transportation, communication, shelter, food, water, medical supplies, and other essentials needed during an evacuation.

People who engage in prepping or emergency preparedness often have a bug-out bag (BOB) prepared, which is a portable kit containing essential items to sustain an individual or a group for a short period during an emergency evacuation. The contents of a bug-out bag may include first aid supplies, water purification tools, non-perishable food, clothing, shelter materials, communication devices (such as radios), and other survival gear.

The decision to "bug out" is often influenced by factors such as the severity of the threat, the level of preparedness of the individual or group, and the availability of resources in the destination area. It's essential for individuals and families to plan and practice bug-out scenarios to ensure a swift and safe response in case of emergencies.

Using radios in a bug-out scenario is a smart and practical way to maintain communication, coordinate movements, and stay informed. Whether you're evacuating due to a natural disaster, civil unrest, or any other emergency situation, having a reliable means of communication is crucial for the safety and well-being of you and your group. Here are some tips for effectively utilizing radios in a bug-out scenario:

Choose the Right Radios:

Select radios that are suitable for your needs. Baofeng radios, for example, are popular for their affordability and versatility. Ensure that the radios have the necessary range, features, and battery life for your bug-out plans.

Frequency Programming:

Pre-program frequencies and channels that you plan to use during the bug-out scenario. Include emergency channels, local weather channels, and any frequencies that you and your group members will use for communication.

Communication Plan:

Develop a communication plan outlining how your group will use the radios. Assign roles and responsibilities, establish signal codes for various situations, and set protocols for reporting important information. Regularly review and practice the communication plan.

Emergency Channels and Weather Updates:

Monitor emergency channels for updates and information relevant to your bug-out situation. Radios with weather band capabilities can provide real-time weather updates, helping you make informed decisions about your route and shelter.

Use Headsets and Accessories:

Headsets with built-in microphones can be useful in maintaining discreet communication, especially in situations where silence may be crucial. Utilize accessories such as external microphones, earpieces, and handsets for convenience and privacy.

Coordinate Movements:

Use radios to coordinate movements and keep your group together. Establish check-in points and times to ensure everyone is accounted for during the evacuation. Radios help in avoiding unnecessary exposure and ensuring a unified response to challenges.

Set Up Rally Points:

Identify and communicate rally points where your group can regroup if separated. Having predefined locations simplifies the process of finding each other during a bug-out scenario.

Maintain OPSEC (Operational Security):

Be mindful of operational security and avoid sharing sensitive information over the airwaves. Use coded language or prearranged signals to communicate without revealing critical details to potential eavesdroppers.

Battery Management:

Conserve battery life by turning off radios when not in use. Bring spare batteries or portable chargers to ensure a continuous power supply. Be aware of the battery status and replace or recharge as needed.

Practice Regularly:

Conduct drills and practice using the radios in a controlled environment to ensure everyone is familiar with their operation. Regular training enhances the efficiency and effectiveness of communication during high-stress situations.

Scanning and Monitoring:

Use scanning features on your radios to monitor multiple channels simultaneously. Stay aware of potential threats, changing conditions, or incoming information that may impact your bug-out plans.

Stay Informed:

Stay tuned to news and emergency broadcasts for updates on the situation. Radios can provide critical information about road closures, evacuation routes, and other factors that may affect your bug-out strategy.

Chapter 7

Advanced Features and Hacks

Baofeng radios, particularly those from the UV-5R series, are known for their affordability and versatility. While they provide a range of features out of the box, there are some advanced features and hacks that enthusiasts and experienced users have discovered. Keep in mind that modifying your radio beyond its intended use may void warranties or violate regulations, so proceed with caution and ensure compliance with local laws. Here are some advanced features and hacks for Baofeng radios:

Frequency Range Expansion:

Some users have explored ways to expand the frequency range of Baofeng radios beyond the limits set by the manufacturer. This might involve using software or firmware modifications, but caution is advised to avoid legal and regulatory issues.

Software Programming:

Utilize programming software to manually program frequencies, channels, and other settings. This can be more efficient than manual programming through the radio's keypad.

CTCSS/DCS Tones:

Experiment with different CTCSS (Continuous Tone-Coded Squelch System) and DCS (Digital-Coded Squelch) tones to enhance privacy and reduce interference. Some users have found creative ways to use specific tones for group communication.

Dual Watch/Dual Reception:

Baofeng radios often have a dual-watch feature, allowing you to monitor two frequencies simultaneously. This can be useful for staying informed about multiple channels or frequencies.

Power Output Adjustment:

Some users have explored ways to adjust the power output of Baofeng radios, although this may be limited by regulatory requirements. Reducing power output can help conserve battery life.

VFO/MR Mode:

Baofeng radios typically have both VFO (Variable Frequency Oscillator) and MR (Memory Recall) modes. VFO mode allows tuning to any frequency within the radio's range, while MR mode accesses pre-programmed memory channels.

Cross-Band Repeater Operation:

Some Baofeng radios support cross-band repeater functionality, allowing them to act as a repeater between two different frequency bands. This can extend the effective communication range.

Voice Scrambling:

Some Baofeng radios have a voice scrambling feature that can be used to add a layer of privacy to your communication. However, keep in mind that this is not secure encryption and can be easily deciphered by those with the right equipment.

Custom Antennas:

Experimenting with different antennas can affect the radio's performance. Some users have tried using aftermarket antennas to improve signal strength and reception.

External Microphones and Accessories:

Explore the use of external microphones, headsets, and other accessories to enhance the functionality and convenience of your Baofeng radio.

Custom Chirp Firmware: Install Chirp, a free software that unlocks customization options like adjusting power output, setting custom beep tones, and even enabling hidden features like crossband repeat.

DIY Modifications: For the tech-savvy, antenna upgrades and internal tweaks can further enhance your Baofeng's performance. However, proceed with caution and research thoroughly to avoid damaging your radio.

Customizing Baofeng Radios for Specific Needs

Customizing Baofeng radios for specific needs can enhance their functionality and adapt them to particular use cases. While it's essential to ensure compliance with local radio regulations and laws, there are several customization options that users can explore:

Programming Frequencies and Channels:

Utilize programming software to manually program specific frequencies and channels relevant to your needs. This is especially useful for users operating in specific frequency bands or for designated purposes.

CTCSS/DCS Tones:

Customize CTCSS (Continuous Tone-Coded Squelch System) or DCS (Digital-Coded Squelch) tones for privacy and to reduce interference. Assigning specific tones to different groups within your organization can help streamline communication.

Naming Channels:

Assign names or labels to programmed channels for easy identification. This is particularly helpful when managing a large number of channels or frequencies.

Power Output Adjustment:

Depending on your specific use case, you may adjust the power output of the radio. Reducing power output can conserve battery life, while increasing it may be necessary for extended communication range.

Dual Watch/Dual Reception:

Take advantage of the dual-watch feature to monitor two frequencies simultaneously. This can be useful in scenarios where information is broadcast on different channels.

Scanning Preferences:

Customize scanning preferences based on your needs. You can set the radio to scan specific frequency ranges, skip unwanted channels, or adjust the scanning speed.

Voice Scrambling:

If your communication requires an additional layer of privacy, experiment with the voice scrambling feature. Keep in mind that this is not secure encryption, but it can deter casual eavesdropping.

Cross-Band Repeater Operation:

If your Baofeng model supports cross-band repeater functionality, customize it to extend communication range in specific scenarios. This feature allows the radio to act as a repeater between two different frequency bands.

Emergency Alerts and Alarms:

Customize emergency alerts or alarms to draw attention to critical messages. Some Baofeng radios have features that allow users to send and receive emergency signals.

Custom Antennas:

Experiment with different antennas based on your communication needs. For example, a longer antenna may improve signal strength and range, while a shorter one may offer more portability.

External Microphones and Accessories:

Customize your setup with external microphones, headsets, and other accessories. This can enhance the convenience and comfort of using the radio in specific situations.

Backlight and Display Settings:

Adjust the backlight and display settings for optimal visibility in different lighting conditions. This can be particularly useful during nighttime operations.

Before making any customizations, refer to the user manual for your specific Baofeng model to understand its features and limitations. Additionally, stay informed about local regulations and laws governing the use of radios and ensure that your customizations comply with these requirements.

BONUS PAGE

Baofeng Model Comparison Chart

Baofeng UV-5R:
Frequency Range: 136-174 MHz (VHF), 400-520 MHz (UHF)
Power Output: 4W/1W
Channels: 128 programmable channels
Features: Dual-band, dual display, dual standby, FM radio, LED flashlight

---

Baofeng UV-5RTP:
Frequency Range: 136-174 MHz (VHF), 400-520 MHz (UHF)
Power Output: 8W/4W/1W
Channels: 128 programmable channels
Features: Tri-Power (8W/4W/1W), dual-band, dual display, dual standby, FM radio, LED flashlight

---

Baofeng UV-82:
Frequency Range: 136-174 MHz (VHF), 400-520 MHz (UHF)
Power Output: 5W/1W
Channels: 128 programmable channels
Features: Dual-band, dual display, dual standby, FM radio, LED flashlight

Baofeng BF-F8HP:
Frequency Range: 136-174 MHz (VHF), 400-520 MHz (UHF)
Power Output: 8W/4W/1W
Channels: 128 programmable channels
Features: Tri-Power (8W/4W/1W), dual-band, dual display, dual standby, FM radio, LED flashlight

Baofeng UV-9R:
Frequency Range: 136-174 MHz (VHF), 400-520 MHz (UHF)
Power Output: 8W/4W/1W
Channels: 128 programmable channels
Features: Tri-Power (8W/4W/1W), dual-band, dual display, dual standby, FM radio, LED flashlight, IP67 waterproof and dustproof.

Baofeng DM-5R:
Frequency Range: 136-174 MHz (VHF), 400-520 MHz (UHF)
Power Output: 5W/1W
Channels: Digital and analog modes, supports DMR (Digital Mobile Radio)
Features: Dual-band, dual display, dual standby, FM radio, LED flashlight

Baofeng UV-5X3:

Frequency Range: 136-174 MHz (VHF), 400-520 MHz (UHF), 220-225 MHz (220 Ham Band)

Power Output: 5W/1W

Channels: 128 programmable channels

Features: Tri-Band, dual display, dual standby, FM radio, LED flashlight

The above model comparison chart helps you to gain clarity and confidence in your Baofeng radio selection. In a single glance, you have assess to key specifications, advanced features, and user insights across various models, ensuring you choose the radio that perfectly aligns with your communication needs. Discover hidden functionalities, and navigate legal considerations effortlessly. This concise chart streamlines your decision-making process, empowering you to make the most informed choice and unlock the full potential of your Baofeng radio for any situation.